X-TREME FACTS: SCIENCE

WEATHER

by Catherine C. Finan

BEARPORT PUBLISHING

Minneapolis, Minnesota

Credits:

Cover and title page, © solarseven/Shutterstock, © viper345/Shutterstock, and © Riddick Patrec/Shutterstock; 4 left, Anut21ng Photo/Shutterstock; 4, Andriy Blokhin/Shutterstock; 5 top, Caleb Holder/Shutterstock; 5 top left, Photographee.eu/Shutterstock; 5 middle, Heide Pinkall/Shutterstock; 5 bottom left, iofoto/Shutterstock; 5 bottom right, NicoElNino/Shutterstock; 5 bottom left center, Traveller70/Shutterstock; 5 bottom right center, Earth Trotter Photography/Shutterstock; 6 top left, Aleksandr Simonov/Shutterstock; 6 top right, Alexander Canas Arango/Shutterstock; 6 bottom, Aliaksei Baturytski/Shutterstock; 6 bottom left, Slatan/Shutterstock; 6 bottom right, Oleinik Iuliia/Shutterstock; 7 top, Andrey Skutin/Shutterstock; 7 top left, Paul Michael Hughes/Shutterstock; 7 top right, Syda Productions/Shutterstock; 7 middle, prapann/Shutterstock; 7 bottom, Jaromir Chalabala/Shutterstock; 7 bottom left, Zapylaiev Kostiantyn/Shutterstock; 8 bottom, Stock-Asso/Shutterstock; 8 bottom right, AnnaTamila/Shutterstock; 9 top, Adwo/Shutterstock; 9 top left, Kues/Shutterstock; 9 top right, Elnur/Shutterstock; 9 middle, B.navez/Creative Commons; ; 9 bottom, 10 bottom, 23 bottom middle, 24, Joshua Avramson; 9 top, Nabin96/Shutterstock; 11 top, Globe Newspaper Co./Public Domain; 11 top right, Public Domain; 11 middle, Lynn Yeh/Shutterstock; 11 bottom, U.S. National Archives and Records Administration/Public Domain; 11 bottom left, Arthur Rothstein, for the Farm Security Administration/Public Domain; 12 top, Zenobillis/Shutterstock; 12 bottom, FotoKina/Shutterstock; 12 bottom left, chaiyapruek youprasert/Shutterstock; 13 top, MDay Photography/Shutterstock; 13 top second from right, Dejan-Milosavljevic/Shutterstock; 13 top middle, Gabriel Pahontu/Shutterstock; 13 top right, khlongwangchao/Shutterstock; 13 bottom, ymphotos/Shutterstock; 14 top, 14 bottom, Cammie Czuchnicki/Shutterstock; 14 bottom left, V_Lisovoy/Shutterstock; 14 bottom right, WilleeCole Photography/Shutterstock; 15 top left, Jim.henderson/Creative Commons; 15 top middle, Regine Poirier/Shutterstock; 15 top, George Munger/Public Domain; 15 bottom, Rasica/Shutterstock; 15 bottom center, CoolimagesCo/Shutterstock; 15 bottom left, Rommel Canlas/Shutterstock; 15 bottom middle, 16 bottom right, 23 top left, Milkovasa/Shutterstock; 15 bottom right, glenda/Shutterstock; 16 top, Amanda Carden/Shutterstock; 16, Rob Atherton/Shutterstock; 17 top, Dan Borsum, NOAA/NWS/WR/WFO/Billings Montana/Public Domain; 17 middle, Andy Dean Photography/Shutterstock; 17 bottom, Bureau of Land Management/Public Domain; 18 top, justoomm/Shutterstock; 18 top right, Joe Tabacca/Shutterstock; 18 top left, photoschmidt/Shutterstock; 18 bottom, VedranKrkic/Shutterstock; 18 bottom middle, Peter Gudella/Shutterstock; 19 top, NOAA Weather in Focus Photo Contest 2015/Public Domain; 19 top left, 22 bottom right, LightField Studios/Shutterstock; 19 bottom, S.Z/Shutterstock; 19 bottom middle, Erik Lam/Shutterstock; 19 bottom right, Kabardins photo/Shutterstock; 20, Vasin Lee/Shutterstock; 20 bottom right, Twinsterphoto/Shutterstock; 21 top, TTstudio/Shutterstock; 21 bottom, Hedley Lamarr/Shutterstock; 22 top, National Weather Service-Wichita, Kansas/Public Domain; 22 middle, Ben Romalis/Shutterstock; 22 bottom left, Kues/Shutterstock; 23 top, Ikebana Art-studio/Shutterstock; 23 top right, Ljupco Smokovski/Shutterstock; 23 bottom left, right, Oqbas/Shutterstock; 25 top middle, Natykach Nataliia/Shutterstock; 25 top middle lower, Paul Looyen/Shutterstock; 25 top right, Asier Romero/Shutterstock; 25 top left, airdone/Shutterstock; 25 top upper right, Gluiki/Shutterstock; 25 middle left, Picture Partners/Shutterstock; 25 middle middle, Collin Smit/Shutterstock; 25 middle right, sweet marshmallow/Shutterstock; 25 bottom, Stanisic Vladimir/Shutterstock; 25 bottom left, xpixel/Shutterstock; 26 top, Piyaset/Shutterstock; 26 top left, Artem Petreev/Shutterstock; 26 bottom, FloridaStock/Shutterstock; 27 top, Angel DiBilio/Shutterstock; 27 bottom, GMC Photopress/Shutterstock; 27 bottom left, DavideAngelini/Shutterstock; 27 bottom right, Fedor Selivanov/Shutterstock; 28 top left, cbies/Shutterstock; 28 top right, Susan Schmitz/Shutterstock; 28 bottom, ssuaphotos/Shutterstock; 28-29, Austen Photography

President: Jen Jenson
Director of Product Development: Spencer Brinker
Editor: Allison Juda

Produced for Bearport Publishing by BlueApple*Works* Inc.
Managing Editor for BlueApple*Works*: Melissa McClellan
Art Director: T.J. Choleva
Photo Research: Jane Reid

Library of Congress Cataloging-in-Publication Data

Names: Finan, Catherine C., 1972- author.
Title: Weather / Catherine C. Finan.
Description: Minneapolis, Minnesota : Bearport Publishing Company, [2021] | Series: X-treme facts: science | Includes bibliographical references and index.
Identifiers: LCCN 2020039260 (print) | LCCN 2020039261 (ebook) | ISBN 9781647476786 (library binding) | ISBN 9781647476854 (paperback) | ISBN 9781647476922 (ebook)
Subjects: LCSH: Weather--Juvenile literature.
Classification: LCC QC981.3 .F53 2021 (print) | LCC QC981.3 (ebook) | DDC 551.5--dc23
LC record available at https://lccn.loc.gov/2020039260
LC ebook record available at https://lccn.loc.gov/2020039261

For more information, write to Bearport Publishing, 5357 Penn Avenue South, Minneapolis, MN 55419. Printed in the United States of America.

Contents

Weather Goes Wild!

At any given moment, someone on Earth is experiencing some seriously wild weather! From ridiculous rains and blowing blizzards to wicked wind and horrible hail, weather can change in an instant. And although it affects just about everything in our lives, we can't control it. So, whatever the weather brings for temperature, **precipitation**, or wind speed, get ready for a turn toward the extreme!

Raindrops fall at almost 20 miles per hour (32 kph)!

Every minute, about 1,800 thunderstorms are popping up around the planet.

Dirt mixes with winds to create dust storms called **black blizzards.**

Some dust storms have swallowed up entire cities!

Heat waves can turn grapes into raisins right on the vine!

Temperatures in Death Valley, California, can reach over 130 degrees Fahrenheit (54 degrees Celsius). In Antarctica they can drop below −140°F (−96°C).

Look to the Clouds

Whatever the weather, clouds have something to do with it. But how do these wonders of weather—from white and fluffy to dark and dangerous—form? It has to do with water! The sun heats Earth's water, turning it to **vapor** that rises into the **atmosphere**. There, it cools and **condenses**, forming tiny water droplets that come together as clouds. Different types of clouds bring different kinds of weather. Look up—what do you see?

Cirrus clouds are made of ice crystals. They form high in the atmosphere and are a sign of fair weather.

Cumulus clouds may look light and fluffy, but they **can weigh over a million pounds (450,000 kg)!**

Stratus clouds give us gray, overcast days—and maybe a little drizzle, too!

During a thunderstorm, **billions—or even trillions—of pounds of water can be packed into one small area of clouds!**

Severe thunderstorms are most common in the spring and summer.

Some cloud formations look like objects or even animals!

You've probably walked through clouds. Fog is a type of cloud along the ground.

Here Comes the Rain

What happens after the sun's heat sends water to the sky? What goes up must come down! When clouds form and fill up with enough water droplets, the water falls as rain, snow, sleet, or hail. When rain clouds roll in, they can mess up our plans! Though it's not fun to sit inside on a rainy summer day, where else would all the water in the clouds go?

Love the smell of a rain shower? It comes from rainwater mixing with bacteria in the soil.

About one billion tons (907 million MT) of rain fall on Earth every minute!

The Atacama Desert in Chile has areas where rainfall has never been recorded!

Thanks to some **algae** growing nearby, rain the color of blood fell in Kerala, India, in July 2001. Freaky!

Mawsynram, India, gets enough rain every year to cover a four-story building.

Fear the Flood, Dread the Drought

Rain is essential to life on Earth. But when a lot of rain falls in a short amount of time, it can cause a flood. A **drought** can be equally dangerous. Whether washed away in water or dried up in a drought, too much or too little water can cause chaos for crops, animals, and people.

One of the deadliest floods in history **covered an area of land larger than England!**

Walking or driving through a flood can be very dangerous. **Always follow the National Weather Service's warnings.**

In a flash flood, walls of water 30 feet (9 m) high can suddenly sweep over an area.

One of history's weirdest floods wasn't caused by water. The Great **Molasses** Flood of 1919 rushed through the streets of Boston.

The Great Molasses Flood was set off after a molasses storage tank burst open.

Some droughts have lasted more than 10 years!

During the 1930s, **droughts across North America caused huge dust storms.**

Insane Hurricane

Some of the worst floods in history have been caused by hurricanes. These monster storms can form only under just the right conditions, but they are sure to bring extreme weather when they do. Near the **equator**, the sun's heat warms large areas of the ocean, making it so lots of water can **evaporate**. When this rising air is helped along by Earth's **rotation**, a storm can spin and grow stronger, eventually becoming a hurricane.

The weather at the very center of a hurricane can actually be calm and sunny!

Hurricanes, typhoons, and cyclones are all watery weather events. That's because they **are the same thing!**

Florida is hit by about 40 percent of all the hurricanes that strike land in the United States.

Many people from around the country came with boats to help rescue people and animals trapped by Harvey's floodwaters.

Torn Apart by a Tornado

Hurricanes aren't the only storms caused by wicked winds. Tornadoes are some of the most terrifying weather events on the planet—and some of the most damaging. Most tornadoes start from swirling thunderstorm clouds called supercells. Under the right conditions, a spinning **funnel**-shaped part of the supercell cloud reaches toward the ground. A tornado's rotating winds can reach a speed of over 300 mph (480 kph), leaving a trail of destruction behind!

A **tornado watch** means a tornado might form. A **tornado warning** means a tornado has been spotted!

Dogs and cats seem to know when a tornado is coming!

In 1814, a tornado struck Washington, D.C. It drove out British **invaders** who had set fire to the White House the day before!

Only a burned-out shell of the White House remained, but Americans got right to work rebuilding it.

Spinning Fire and Whirling Water

Tornadoes are freaky enough, but what about tornadoes of fire or water? Fire tornadoes can happen in the intense heat of wildfires. They are dangerous columns of spinning fire. Waterspouts are tornadoes over water. While fair-weather waterspouts form over water, tornadic waterspouts start on land before moving to water

Waterspouts can be up to 330 ft (100 m) wide and **last for an hour!**

Florida has more waterspouts than any other place in the United States. So much for the Sunshine State!

Fire tornadoes can be as tall as a 40-story building!

Fire tornadoes are also called fire whirls, firenadoes, fire twisters, and even fire devils!

The temperature inside a firenado can reach 2,000°F (1,090°C)!

Firefighters use special drones to search for wildfire hotspots before they turn into firenadoes.

Blinded by a Blizzard

We know that wind and water can be a nasty mix. When that water is in the form of snow, it can get even nastier! It's fun to go outside to make a snowman or go sledding. But when snow is whipped by fierce winds, things get a little . . . less fun. Combine snow with wind that reaches 35 mph (56 kph) or more and you have a blizzard. It's time to stay inside with your hot cocoa!

A blizzard can have winds that are as strong as hurricane winds.

In 2010, a massive blizzard nicknamed **Snowmageddon brought historic snowfalls to the mid-Atlantic states.**

Lake-effect snow happens when cold air moves over a warm lake, creating clouds that dump lots of snow very quickly!

The lake-effect snow that hit Buffalo, New York, in 2014 dumped between 5 and 7 feet (1.5 and 2.1 m) of snow in some neighborhoods.

Lake-effect snow might pile up lots of snow in one place but almost nothing just a mile (1.6 km) away!

All Lit Up

Depending on where you live, you might never be blown around by a blizzard. But chances are good that you've been rattled and shaken by a storm with thunder and lightning. Water droplets and ice crystals moving around inside storm clouds can create electricity. The lightning we see is actually the spark of this electricity. Let's check out some other enlightening facts about lightning!

Lightning's extreme heat makes the air around it expand, which creates the sound of thunder.

Lightning can be five times hotter than the surface of the sun!

The average flash of lightning has enough energy to keep a 100-watt light bulb lit for three months!

The Empire State Building in New York City is struck by lightning about 25 times a year!

Lightning usually strikes the tallest objects around. So, if you're caught in a thunderstorm, never hide under tall trees.

About 100 bolts of lightning hit Earth every second!

Hard-Hitting Hail

Thunderstorms don't just bring flashing lightning, roaring thunder, and heavy rain. Sometimes they also pelt the ground with chunks of ice. *Ouch!* Hail forms when water droplets freeze together in the very cold upper part of a storm cloud. These frozen droplets are tossed around in the cloud and grow larger as layers of ice form around them. When the hail finally falls from the cloud, you'd better get inside quickly!

Hail can be the size of a softball and come cruising to the ground at 100 mph (161 kph).

In 2018, people in Argentina found a **hailstone the size of a soccer ball!**

In the 1700s, Europeans rang church bells and fired cannons toward the sky in an effort to prevent hail!

Hail falls so often on the Great Plains region of North America that it's sometimes called Hail Alley.

In 1360, a hailstorm near Paris, France, killed hundreds of invading British soldiers.

It's Raining Cats and . . . Frogs?

Getting pelted by big chunks of ice is bad enough. But imagine going for a walk, minding your own business, when a shower of frogs rains down from the sky. That's just hopping crazy! Yet, it's possible. Animal rain happens when smaller animals, such as fish, frogs, birds, and even snakes—*yikes!*—get swept up in waterspouts or rising winds. They fall back to Earth along with rain. Reports of animal rain go back hundreds of years. Better grab a strong umbrella!

A rain of fish was said to have fallen in Europe in 989 CE!

In February 2010, **hundreds of fish fell from the sky in Lajamanu, Australia.**

Spiders falling from the sky sounds like a nightmare, but it happened in Brazil when large webs were scattered by rain and wind.

Eels have rained down on Alabama, **and snakes have fallen from the sky** in Tennessee. Now that's wet and wild!

England is known to be a rainy place. It's even **had a rainfall of snails!**

In June 2005, thousands of tiny frogs rained down upon a small town in Serbia.

The Future of Weather

Our world is full of weather extremes. But we aren't just living with wild weather—we are part of the reason it happens. Earth's **climate** is changing, and our activities that pollute the planet are a major cause. They have led to increased **global warming** and may also be connected to the strength and **frequency** of severe weather. It's up to us to take care of the planet. The future of our weather—and life on Earth—depends upon it!

Earth is warming about 10 times faster than it has in the past.

Warming temperatures in the oceans are destroying underwater life and may lead to stronger hurricanes.

Frogs tell us the climate is changing! They are some of the first animals to die when their habitat changes.

Scientists have suggested shooting mirrors or salt into our atmosphere to help bounce the sun's heat away from Earth to slow climate change.

Some glaciers are wrapped in blankets and foam to keep them from melting due to global warming.

Wild Wind Sock

Craft Project

A wind sock is a tool used to determine the direction and strength of wind. It looks similar to a giant sock, which is where the name came from. Wind socks are commonly found at airports and weather stations. There's a lot you can learn by looking at a wind sock.

What You Will Need

- 2 plastic bags
- Scissors
- Twine
- 4 paper clip
- Tacky glue

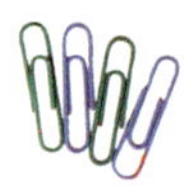

Wind socks are commonly found at airports. They help pilots take off and land safely in all kinds of weather.

Step One

Cut the handles and bottom off a plastic bag so you have a tube shape that is open at both ends.

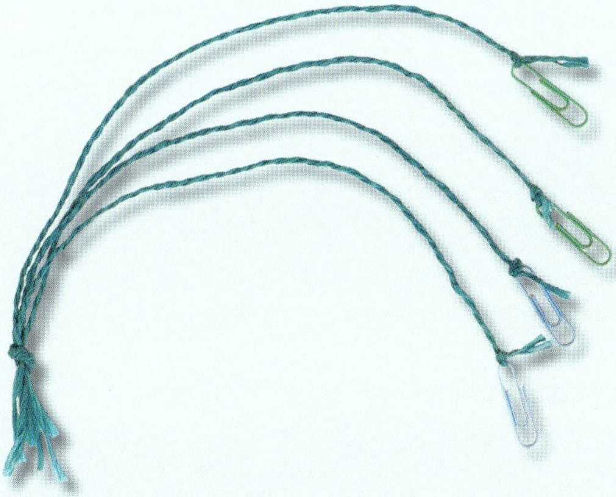

Step Two

Cut four pieces of twine about 14 inches (35 cm) long. Set out four paper clips. Tie a paper clip to the end of each piece of twine. Gather the loose ends of twine and tie them together in a knot.

Step Three

Attach one of the paper clips to the bag by poking the tip of the clip through the bag as shown. Repeat with the other three paper clips, spacing them evenly around one end of the bag.

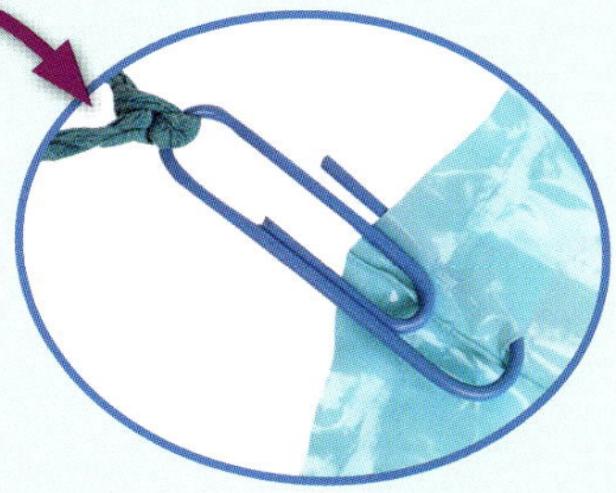

Step Four

Cut the second plastic bag into strips.

Step Five

Glue the plastic strips to the inside of the non-clipped open end of the bag. Let the glue dry. Then, hang your wind sock outside. Watch your wind sock. What can you learn about the wind in your part of the world?

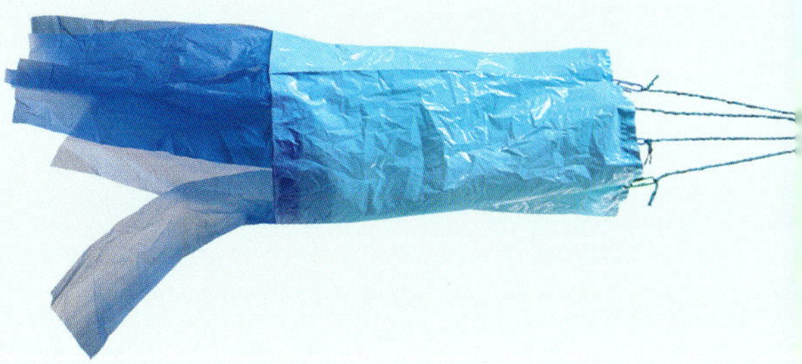

Glossary

algae plantlike living things that are mainly found in water and can make their own food

atmosphere the gases surrounding Earth

climate patterns of weather over a long period of time

condenses changes from a gas into a liquid

drought a long period of time when there is little or no rain

equator the imaginary line around the middle of Earth that is halfway between the North and South Poles

evaporate to change from a liquid into a gas

frequency the number of times that something happens or is repeated in a period of time

funnel an open cone with a tube at the end

global warming a rise in temperatures around the world

habitat a place in nature where a plant or animal normally lives

hemisphere one half of Earth

invaders people or animals that enter a place where they don't normally belong

molasses a thick, sweet, dark-brown syrup made from sugarcane

precipitation water in the form of rain, snow, sleet, or hail

rotation the act of turning on or around an axis

vapor tiny particles of a liquid

Read More

Kenney, Karen Latchana. *Meteorologist (STEM Careers)*. Minneapolis: Jump!, 2019.

Krajnik, Elizabeth. *Clouds and Precipitation (Spotlight on Weather and Natural Disasters)*. New York: PowerKids Press, 2019.

McGregor, Harriet. *Heat Wave Horror! (Uncharted: Stories of Survival)*. Minneapolis: Bearport Publishing, 2021.

Vail, Grace. *More Freaky Weather Stories (Freaky True Science)*. New York: Gareth Stevens Publishing, 2020.

Learn More Online

1. Go to **www.factsurfer.com**
2. Enter "**X-treme Weather**" into the search box.
3. Click on the cover of this book to see a list of websites.

Index

About the Author

Catherine C. Finan is a writer living in northeastern Pennsylvania, where she has experienced her fair share of blizzards. When a snowstorm hits, she curls up with a good book next to the fireplace and watches it from *inside* the house!